Mathematics in a World of Data

DATA-DRIVEN MATHEMATICS

Jack Burrill, Miriam Clifford, Emily Errthum, Henry Kranendonk,
Maria Mastromatteo, and Vince O'Connor

Dale Seymour Publications®
White Plains, New York

This material was produced as a part of the American Statistical Association's Project "A Data-Driven Curriculum Strand for High School" with funding through the National Science Foundation, Grant #MDR-9054648. Any opinions, findings, conclusions, or recommendations expressed in this publication are those of the authors and do not necessarily reflect the views of the National Science Foundation.

Managing Editor: Alan MacDonell

Senior Mathematics Editor: Nancy R. Anderson

Consulting Editor: Maureen Laude

Production/Manufacturing Director: Janet Yearian

Production/Manufacturing Manager: Karen Edmon

Production Coordinator: Roxanne Knoll

Design Manager: Jeff Kelly

Cover and Text Design: Christy Butterfield

Cover Photo: Garry Gay, Image Bank

This book is published by Dale Seymour Publications®, an imprint of Addison Wesley Longman, Inc.

Dale Seymour Publications
10 Bank Street
White Plains, NY 10602
Customer Service: 800-872-1100

Printed in the United States of America.

Order number DS21167

ISBN 1-57232-402-3

1 2 3 4 5 6 7 8 9 10-ML-03 02 01 00 99 98

This Book Is Printed
On Recycled Paper

Authors

Jack Burrill
National Center for Mathematics
Sciences Education
University of Wisconsin-Madison
Madison, Wisconsin

Miriam Clifford
Nicolet High School
Glendale, Wisconsin

Emily Errthum
Homestead High School
Mequon, Wisconsin

Henry Kranendonk
Rufus King High School
Milwaukee, Wisconsin

Maria Mastromatteo
Brown Middle School
Ravenna, Ohio

Vince O'Connor
Milwaukee Public Schools
Milwaukee, Wisconsin

Consultants

Wanda Bussey
Rufus King High School
Milwaukee, Wisconsin

Patrick W. Hopfensperger
Homestead High School
Mequon, Wisconsin

Barbara Perry
R.W. Johnson Pharmaceutical
    Research Institute
Raritan, New Jersey

Jeffrey Witmer
Oberlin College
Oberlin, Ohio

Data-Driven Mathematics Leadership Team

Gail F. Burrill
National Center for Mathematics
Sciences Education
University of Wisconsin-Madison
Madison, Wisconsin

Miriam Clifford
Nicolet High School
Glendale, Wisconsin

James M. Landwehr
Bell Laboratories
Lucent Technologies
Murray Hill, New Jersey

Richard Scheaffer
University of Florida
Gainesville, Florida

Kenneth Sherrick
Berlin High School
Berlin, Connecticut

Acknowledgments

The authors thank the following people for their assistance during the preparation of this module:

The many teachers who reviewed drafts and participated in the field tests of the manuscripts

The members of the *Data-Driven Mathematics* leadership team, the consultants, and the writers

Nancy Kinard, Ron Moreland, Peggy Layton, and Kay Williams for their advice and suggestions in the early stages of the writing

Kenneth Sherrick, Richard Crowe, and Wanda Bussey for their thoughtful and careful review of the early drafts

Kathryn Rowe and Wayne Jones for their help in organizing the field-test process and the Leadership Workshops

Barbara Shannon for her many hours of word processing and secretarial services

Jean Moon for her advice on how to improve the field-test process

Amy Plant for writing answers for the Teacher's Edition

The many students and teachers from Brown Middle School, Nicolet High School, Whitnall High School, Rufus King High School, Homestead High School, and the University of Florida, Gainesville, who helped shape the ideas as they were being developed

Table of Contents

About *Data-Driven Mathematics*

Historically, the purposes of secondary-school mathematics have been to provide students with opportunities to acquire the mathematical knowledge needed for daily life and effective citizenship, to prepare students for the workforce, and to prepare students for postsecondary education. In order to accomplish these purposes today, students must be able to analyze, interpret, and communicate information from data.

Data-Driven Mathematics is a series of modules meant to complement a mathematics curriculum in the process of reform. The modules offer materials that integrate data analysis with secondary mathematics courses. Using these materials will help teachers motivate, develop, and reinforce concepts taught in current texts. The materials incorporate major concepts from data analysis to provide realistic situations for the development of mathematical knowledge and realistic opportunities for practice. The extensive use of real data provides opportunities for students to engage in meaningful mathematics. The use of real-world examples increases student motivation and provides opportunities to apply the mathematics taught in secondary school.

The project, funded by the National Science Foundation, included writing and field testing the modules, and holding conferences for teachers to introduce them to the materials and to seek their input on the form and direction of the modules. The modules are the result of a collaboration between statisticians and teachers who have agreed on statistical concepts most important for students to know and the relationship of these concepts to the secondary mathematics curriculum.

Using This Module

Why the Content Is Important

Numbers are everywhere. They are used to count, sort, rank, or summarize almost anything. Numbers are used to explain what the chances of rain are for today, which school has the best basketball team, and where you can buy the cheapest gas.

The lessons and projects in this module will help you learn how to make sense out of numbers in context, which are called *data*. You will also learn how to communicate effectively with data. By looking at information presented in the media and discovering patterns in activities and simple experiments, you will begin to understand the need for quantitative information in decision making. You will also see how data shape the world in which you live.

The fourteen lessons in this module are divided into four units, which are briefly described here.

Have you ever thought about how decisions are made about which United States cities are the best places to live? How much money do you think people in different regions of the country spend on soft drinks each year? Data can be collected and studied to answer these questions. Unit I of this module includes a variety of mathematical tools for working with data: organizing data in tables or charts, creating plots or graphs, calculating averages, and studying the variability in data.

Data come in many forms, and some, such as the colors of M&Ms® in a package, are not numbers at all. Categorical data and measurement data are explained in Lesson 4 of Unit II. They are summarized in different ways. Data are often used to answer questions and support arguments. An investigation process is developed in Lesson 5 of this unit to provide a structured approach to problem solving. The investigation process is used in Lesson 6 to help you sort, count, and represent data involving M&Ms.

Percents are used to compare groups or populations of different sizes. In Unit III, you will compare areas and populations of the continents on 10-by-10 grids that provide a visual representation for percent. Percents are used in reporting the results of a

M&Ms is a registered trademark of M&Ms/Mars, a division of Mars, Inc.

telephone poll and interpreting data on a shot chart for high-school basketball players. In Lesson 11, percents are also used to summarize data collected by tossing paper cups.

Measurement data are used to measure observations, results, or outcomes. Measurement data can be represented on the real-number line and can be summarized by finding a center. All of the activities and experiments in Unit IV use measurement data. The experiments include measuring your reaction time and finding the average size of bubbles you can blow with a straw.

Each lesson generally begins with an introduction of what is to be presented. Following the introduction, there is an opening section that can be discussed by the class, allowing you to share your experiences. This is followed by a section of problems that includes material that can be completed by the class as a whole, by small groups, or by individual students. Finally, there is a section that includes work to be done individually or in groups.

Content

Mathematics content: You will be able to

- Count, sort, order, and compare rational numbers.
- Compute with positive rational numbers.
- Compute with percent.
- Work with measurement.
- Understand concepts of inequality.
- Understand concepts of probability.
- Use estimation techniques.

Statistics content: You will be able to

- Work with categorical and measurement data.
- Collect data by observation, survey, and experiments.
- Construct frequency and cumulative frequency distributions.
- Represent data in tables, bar graphs, number-line plots, stem-and-leaf plots, scatter plots, and box plots.
- Make predictions.
- Understand concepts of variability, bias, and association.

Technology

The amount of technology you can use may vary. Scientific calculators are essential in some lessons. Graphing calculators or computers with spreadsheet and graphing capabilities could be used in some lessons.

Context and Units

Numbers in Context

Would a score of 95 on a test be good? How about 95 as a golf score? Would you like to score 95 in bowling?

Is a 95-page book long? A 95-page homework assignment?

The number 95 needs a *context* to give it meaning. Almost everything tends to be numbered in one way or another in our society—everything that can be counted, measured, averaged, estimated, or otherwise *quantified*. Calculators and computers help us do amazing things with numbers. But to have a precise meaning in our world, a number must be used in context. Often a number has a unit in its context. Numbers used in context are called *data*.

OBJECTIVES

Recognize units for numbers in context.

INVESTIGATE

The prices a gas station charges for gasoline are usually written on a sign large enough for drivers to see from the road. This allows customers to compare prices. Numbers such as 112^9 are often used to indicate price. In this case, the unit is cents per gallon and the price for one gallon of gasoline is 112.9 cents or 1 dollar and 12.9 cents. What other numbers may be displayed for consumers to make decisions about what to buy or where to shop?

Discussion and Practice

1. Match each item with an appropriate context.

Item	Context
9 × 12	test score
$6\frac{1}{2}$	locker combination
21–7	room size
7–21	shoe size
102–95	temperature
95	population of the United States
75	time
55	cost of a car
20,000	score of a football game
254,000,000	score of a basketball game
2–30–38	miles per hour
2:05	July 21

2. When is the number 6 a good result? Compile a list of situations or contexts in which 6 is a good result. Be sure to indicate the units in each case.

3. When is the number 6 *not* a good result? Compile a list of situations or contexts in which 6 is not a good result. Be sure to indicate the units in each case.

Practice and Applications

4. Below are two sets of data.

$10.07	$12.74
8.687 gallons	10.204 gallons
Unleaded Regular: 115^9	Unleaded Super: 124^9

   a. Describe the probable setting or context in which you would find these numbers.

   b. What would be an appropriate unit for 115^9 and 124^9 in this context?

5. For each unit of data, describe two contexts, one in which 50 is large and one in which 50 is small.

   a. 50 people   b. $50   c. 50 mph   d. 50 minutes

6. Ages of young children are often written with months as the unit. Express the age 22 months in two different ways.

Use the table below for Problems 7–11.

Average Life Expectancy for Certain Animals

Animal	Estimate of Life Expectancy	Actual Life Expectancy
Dog	——————	——————
Hamster	——————	——————
Duck	——————	——————
Horse	——————	——————
Hippopotamus	——————	——————

7. Estimate the average longevity, or life expectancy, of each animal listed.

8. Use a data source or ask your teacher for information to complete the actual-life-expectancy column in the table.

9. Note the variations in the expected longevity among these animals. Which animal typically lives the longest?

10. Make a scatter plot with your estimates on the horizontal axis and actual life expectancy on the vertical axis of your graph.

    a. How do your estimates compare with the actual life expectancy? Explain.

    b. Compare your estimates with those of another student. Write a sentence that explains how they compare.

11. Sometimes life expectancy is given as a range of time. Do you think this is reasonable? Explain.

12. Some animals live a very long time. Listed below are some special animals and the length of time each lived. Use the information and guess what the unit of measurement is for the age of each animal.

Record-Setting Longevity

Joey, K. Ross's canary	Lived to be 408
Fritzy, mouse in Great Britain	Lived to be 392
Flopsy, L. B. Walker's rabbit	Lived to be 18.896
Snowball, Wall's guinea pig	Lived to be 178.5
Rodney, Rodney Mitchell's rat	Lived to be 7.08

Source: *Guinness Book of World Records,* 1993

Extension

13. There is variability in the traits of animals. Not all animals of the same type have the same height, weight, or life expectancy.

Consider this situation: You are a pet-store owner and Buzz Stone comes back to you complaining that the hamster he purchased 12 months ago has died. When Buzz bought the hamster, he asked you about the life expectancy of hamsters and you told him it was about 2 years. What could you tell him?

Summary

A number can have many meanings. The meaning depends on the *context* and the *unit*. It is not possible to tell whether a number represents something large or small, or good or bad, without a context. Numbers written in context are **data**. A range of numbers may be reported when there is *variability* in data. For example, the life expectancy of a horse is 20–25 years.

Communicating with Data

What facts about your city make it a great place to live?

If you could choose to live in any city in the United States, which one would you choose and why?

According to *Money Magazine,* July, 1997, the United States' top ten metropolitan areas in which to live are:

1. Nashua, New Hampshire
2. Rochester, Minnesota
3. Monmouth, New Jersey
4. Punta Gorda, Florida
5. Portsmouth, New Hampshire
6. Manchester, New Hampshire
7. Madison, Wisconsin
8. San Jose, California
9. Jacksonville, Florida
10. Fort Walton Beach, Florida

OBJECTIVE

Recognize how data are used in quality ratings.

INVESTIGATE

Decisions about which cities in the Unites States are best in which to live may be made and reported in the media by people who have not even visited those cities. The rankings for the cities above were based on data concerning weather, health care, recreation, education, transportation, the arts, crime, housing, and jobs. What information would you use to rank the top ten metropolitan areas in the United States?

Discussion and Practice

1. What factors do you think are most important in determining a quality rating for a city?

2. Do you think information regarding how cities are ranked affects choices people make about where they live? Explain.

3. Describe any common features in the cities listed above.

4. Specify information you think would be important to consider in determining a quality rating for each of the following.

   a. Cars
   c. Airlines
   b. Schools
   d. Baseball stadiums

Consider the following question: "How could you rank the major U.S. airlines?" In *USA Today*, April 12, 1994, information was reported for the major airlines that can be used in a quality rating. The table, which looks like an airline report card, includes these four categories.

On-Time Percent: percent of flights that arrive on time
Mishandled Bags: bags that are lost or damaged beyond normal wear
Bumped Passengers: passengers that miss a flight because of overbooking
Average Age of Aircraft: the average age of the planes owned by an airline

Airline Service Report

Of the nine major U.S. airlines, only three—Southwest, USAir, and TWA—improved in service last year, the Airline Quality Rating survey says. American, Delta, Northwest, America West and Continental got worse. United held steady. Major airlines are defined as having operational revenue of at least $1 billion.

On-Time Percent		Mishandled Bags	
Southwest	89.5%	Southwest	38.1
Northwest	85.9%	America West	44.1
America West	85.5%	TWA	50.5
USAir	82.9%	American	56.9
TWA	82.5%	Delta	57.3
American	80.8%	Northwest	58.9
Continental	79.0%	USAir	59.0
United	78.4%	Continental	61.1
Delta	76.7%	United	64.8

Bumped Passengers		Average Age of Aircraft	
American	0.36	Southwest	7.3 years
United	0.36	America West	7.6 years
USAir	0.68	American	8.9 years
Delta	0.73	Delta	9.1 years
Northwest	1.21	United	10.8 years
TWA	1.58	USAir	11.0 years
Continental	1.69	Continental	15.3 years
America West	2.10	Northwest	16.4 years
Southwest	3.18	TWA	18.7 years

NOTE: Numbers of mishandled bags and bumped passengers are per 10,000 passengers.

Source: U.S. Department of Transportation, *World Aviation Directory*

5. Refer to the airline service report above.

   a. How were the major airlines identified?

   b. The note below the tables is a key for interpreting some of the numbers. Delta Airlines has a mishandled-bag rating of 57.3. Write a sentence explaining what this number means.

   c. The number of bumped passengers for United and American Airlines is 0.36 per 10,000 passengers. If one of these airlines serves a million people in a year, about how many of these would be bumped?

6. Use the information from the table to identify each of the nine major airlines as *excellent, good,* or *fair.*

7. Do you think it is always better for an airline to have newer airplanes? Explain.

8. Read the article *Southwest Ranks Best in Airline Survey* on page 10. It was written in the middle 1990s. Does your quality rating system agree with the article's point of view? Explain.

Southwest Ranks Best in Airline Survey

Southwest Airlines bumped American in 1993 to become the USA's best major airline.

TWA, worst for three years, left that spot to Continental.

The rankings are based on Department of Transportation statistics. The annual survey—done by Wichita State University and the University of Nebraska—considers such things as lost bags, late flights, age of fleet, complaints, and denied boardings.

Overall, service on the USA's nine major airlines worsened for the fourth year. "Airline travel has become a big bus ride," researcher Dean Headley says. U.S. airlines have lost $11.4 billion the past four years. As their finances improve, Headley expects service to rebound.

American had been No. 1 for three years. A flight attendants' strike in November probably caused more complaints.

Southwest had the fewest lost bags and best on-time performance. That makes sense. Most of its routes are short and fliers tend not to check bags. Southwest also avoids congested airports, which cause delays. American had the fewest denied boardings.

TWA showed the most improvement. Last year, it emerged from Chapter 11 bankruptcy reorganization and became 45% employee-owned.

Source: *USA Today*

Practice and Applications

9. The following table lists some of the test data recorded in a rating of some mid-priced sports sedans. Use these data to identify the best and the poorest cars in terms of acceleration, fuel economy, and braking from 60 mph.

Car-Test Data

	Infiniti G20	Volkswagen Jetta III	Acura Integra	Nissan Altima
Acceleration				
0–30 mph, sec	4.0	3.6	3.6	3.3
0–60 mph, sec	10.1	8.9	9.2	8.7
45–65 mph, sec	6.9	5.4	5.5	5.6
Fuel Economy				
EPA, mpg (city/hwy)	24/32	18/25	25/31	24/30
CU's 150-mile trip, mpg	34	27	35	32
City driving, mpg	20	16	21	19
Gallons of fuel, 15,000 mi	515	665	495	650
Cost of fuel, 15,000 mi	$615	$795	$595	$660
Braking from 60 mph				
Dry pavement, ft	136	134	142	132
Wet pavement, ft	149	156	170	149

Source: *Consumer Reports,* November, 1994

Extension

10. Some athletes are models for young people today. In a poll conducted by a Sports Marketing Group, reported in the *Kalamazoo Gazette* on June 23, 1994, the four most important positive character traits of athletes, as identified by 12–17 year olds, were: caring, smart, good sport, and trustworthy. The negative character traits identified by 12–17 year olds were: cheater, cocky, greedy, and weak.

   a. Do you think these qualities should be used to rank athletes as sports heroes? Explain.

   b. Would you consider other positive or negative traits? Which ones?

Summary

Data on climate, education, health care, recreation, and so on, are used to compare and rank cities. Data are also used to rate or rank airlines and cars. In the development of ranks or ratings, information is quantified and decisions are made about what is favorable and what is unfavorable.

Data in Rows and Columns

What is your favorite kind of soda to drink?

Can you describe the advertisements for your favorite soda?

Do you think advertisements have affected your preference?

<table>
<tr><td>

OBJECTIVES

Make sense of data in tables and other data displays.

Find patterns and make comparisons in data.

</td></tr>
</table>

Each percent of the soft-drink market is worth about 460 million dollars a year to a company in retail sales. It is, therefore, not surprising that millions of dollars are spent advertising soft drinks. In 1991, per-capita consumption of soft drinks was 46.7 gallons. In this lesson, you will learn about your own soda-consumption patterns and compare them with the patterns of people in many regions of the nation.

INVESTIGATE

About how much soda do you drink in a day, a week, a month, and a year? How could you calculate an estimate for the amount of soda you drink in a year?

Discussion and Practice

1. Set up a table like the one below, or use *Activity Sheet 1*. Record individual soda-consumption estimates for each person in the class in terms of numbers of 12-ounce cans.

Soda Consumption

Names	Day	Week	Month	Year
_____	_____	_____	_____	_____
_____	_____	_____	_____	_____
Averages	_____	_____	_____	_____

2. Refer to your soda-consumption data.

  a. Make a number-line plot of the data for daily soda consumption in 12-ounce cans.

  b. Describe the variability of the class data for daily soda consumption.

  c. Find the average amount of soda consumed each day by students in the class.

3. What is the estimated average amount of soda consumed per student for your class, in gallons per year?

4. Find out how many students attend your school and use that information for the following problems.

  a. Use your class average to estimate the amount of soda consumed daily by students in your school. Be sure to specify the units.

  b. About how much money is spent on soda each day by students in your school?

  c. If all the students in your school were to drink soda from cans, about how many cans would be thrown away each year?

Data are often organized in tables or charts. When data are organized it may be easier to make comparisons, find patterns, recognize variability, and calculate averages. It is also helpful to represent data in plots or graphs.

Practice and Applications

The estimates in the table that follows were based on a survey of bottlers. Data include seltzer and club soda marketed by soft-drink bottlers but do not include noncarbonated drinks. *Per capita* means "per person."

Yearly Consumption of Soft Drinks in the U.S. by Region

Region	Population (millions)	Total Gallons Consumed (millions)	Per-Capita Consumption (gallons)	Retail Sales (millions of dollars)
New England	12.8	554.5	43.3	$ 2,076.9
Middle Atlantic	37.6	1,737.9	46.2	$ 6,936.5
East North Central	42.1	2,033.3	48.3	$ 7,518.0
West North Central	17.8	868.4	48.8	$ 2,941.8
South Atlantic	42.8	2,152.8	50.2	$ 7,181.0
East South Central	15.4	845.6	54.9	$ 2,862.5
West South Central	27.1	1,303.8	48.1	$ 4,570.5
Mountain	13.4	496.8	37.1	$ 2,013.2
Pacific	37.6	1,538.2	40.9	$ 6,268.8
Totals	246.8	11,531.2	46.7	$43,106.2

Source: *Beverage World,* May, 1990

5. Refer to the table above.

   a. In which region do you live?

   b. How much soda is consumed each year in the region where you live? Write your answer without a decimal point and with appropriate units.

6. In which two regions are the greatest number of gallons of soda consumed? The least?

7. Refer to the per-capita column in the table above.

   a. Identify the regions in which soda consumption is the greatest and the least.

   b. Are these the same regions you identified in Problem 6? Explain why or why not.

   c. What do you observe about the consumption of soda in the U.S.?

8. How much money in dollars is spent on soda each year in the U.S.?

9. How does the U.S. per-capita total compare to the average amount consumed by students in your class in gallons per year?

Extension

10. Explain how the data in the per-capita column are calculated.

11. If you average the data in the per-capita column, will you get 46.4 instead of 46.7? Explain why 46.7 is really the correct answer.

12. Are soft-drink prices the same in different regions of the country? Justify your answer.

Summary

When data are organized in tables, it is usually easier to make comparisons and find patterns. The arrangement of data in tables facilitates the calculation of averages.

Summary for Unit I

- A number can have many meanings. The meaning depends upon the *context* and the *unit*. It is not possible to tell whether a number represents something large or small, good or bad, without a context. Numbers written in context are called *data*. A range of numbers may be reported when there is *variability* in data. For example, the life expectancy of a cat is 10–12 years.

- In the development of ranks or ratings, information is quantified and decisions are made about what is favorable and what is unfavorable. Quantified information may include counts such as number of people, averages of several numbers, percents, and rates such as miles per gallon.

- Data are often organized in tables. When data are organized it may be easier to make comparisons, find patterns, recognize variability, and calculate averages. It is also helpful to represent data in plots or graphs.

Assessment for Unit I

1. List two situations for which 10 is a good result. Specify the units in each case.

2. Specify information you think would be important to consider in determining a quality rating for color televisions.

3. The following chart shows how long things last or the age at which products are replaced in years. Use the data for parts a–f on page 17.

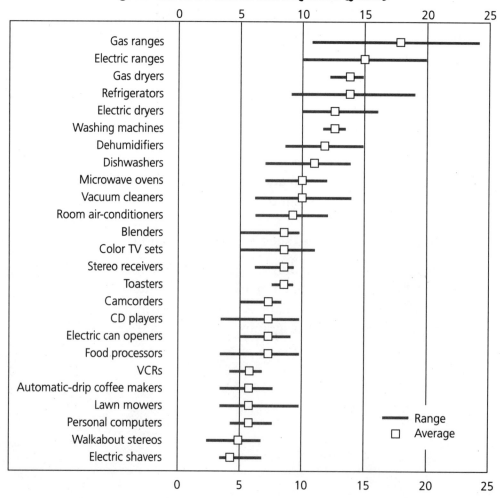

Age at Which Products Are Replaced (years)

Source: *Appliance Magazine,* a Dana Chase publication

a. Which products have to be replaced about every 10 years?

b. About how often does a color television have to be replaced?

c. Which products have to be replaced most often?

d. Which product has the greatest range of years for replacement time? What is the range?

e. About how many new toasters should someone expect to buy in a period of 15 years?

f. Explain why the age a product has to be replaced is given as a range.

4. The following tables describe the weather conditions in Mt. Rainier National Park, about 100 miles from Seattle, Washington, and the Golden Gate National Recreational Area in San Francisco, California. After you study the tables of numbers, write a paragraph for each location describing the weather and what you might expect if you planned to visit these areas.

Mount Rainier National Park

Weather Parameters	Month											
	J	F	M	A	M	J	J	A	S	O	N	D
Temperature												
Normal Daily Maximum	34	38	43	51	59	64	74	75	68	57	42	34
Normal Daily Minimum	12	15	19	25	32	37	42	41	35	29	22	19
Extreme High	55	60	73	80	88	93	101	97	94	85	65	61
Extreme Low	−38	−35	−18	−3	13	22	27	24	18	−5	−12	−16
Days Above 90°	0	0	0	0	0	0	2	2	1	0	0	0
Days Below 32°	30	28	30	27	17	5	1	1	9	21	28	30
Precipitation												
Normal	14.6	10.2	8.9	6.3	4.0	3.8	1.6	2.9	4.6	7.7	12.1	15.9
Maximum	30.4	20.8	19.5	12.5	9.1	8.0	6.0	7.2	15.2	23.6	25.4	29.1
Maximum 24-Hr Precipitation	5.7	4.8	3.3	5.4	2.2	2.8	2.3	2.5	4.5	5.3	7.9	7.9
Maximum Snowfall	193	182	155	107	46	9	6	T	27	65	138	164
Days with Measurable Precip.	22	20	21	19	16	15	9	11	13	16	20	23
Average No. Thunderstorms	0	0	0	0	1	2	1	2	1	0	0	0
Sunshine/Cloudiness												
No. Clear Days	3	2	3	3	4	6	13	12	10	6	3	2
No. Partly Cloudy Days	2	3	4	5	6	5	7	8	6	5	3	2
No. Cloudy Days	26	23	24	22	21	19	11	11	14	20	24	27
% Possible Sunshine	28	35	42	50	52	55	65	65	55	45	30	20

Golden Gate National Recreation Area

Weather Parameters	Month											
	J	**F**	**M**	**A**	**M**	**J**	**J**	**A**	**S**	**O**	**N**	**D**
Temperature												
Normal Daily Maximum	56	59	59	60	61	63	64	65	67	66	63	57
Normal Daily Minimum	43	44	44	44	47	50	51	52	52	49	46	44
Extreme High	74	77	79	84	90	83	81	94	91	91	87	79
Extreme Low	29	30	32	32	37	39	42	41	40	35	30	27
Days Above 90°	0	0	0	0	0	0	0	0	0	0	0	0
Days Below 32°	1	0	0	0	0	0	0	0	0	0	0	0
Precipitation												
Normal	5.3	3.7	3.5	2.1	0.6	0.2	0.1	0.2	0.4	1.6	3.0	4.4
Maximum	11.4	10.8	9.4	7.4	4.1	1.4	1.0	1.6	3.7	11.0	9.5	13.8
Maximum 24-Hr Precipitation	4.9	2.4	2.0	2.3	1.1	1.1	0.6	0.6	2.6	4.9	2.4	4.1
Maximum Snowfall	0	0	0	0	0	0	0	0	0	0	0	0
Days with Measurable Precip.	11	10	10	6	3	1	1	1	2	4	8	10
Average No. Thunderstorms	0	0	0	0	0	0	0	0	0	0	0	0
Sunshine/Cloudiness												
No. Clear Days	9	8	10	11	13	15	17	15	16	14	11	9
No. Partly Cloudy Days	7	7	9	10	11	10	11	12	10	9	8	8
No. Cloudy Days	15	13	12	9	7	5	3	4	4	8	11	14
% Possible Sunshine	56	62	69	73	72	73	66	65	72	70	62	53

Source: *The Complete Guide to America's National Parks,* 1990–1991

Sorting and Counting

Categorical and Measurement Data

How are movies rated for different audiences?

How much money does it cost to make a movie?

You have been using data that are numbers in context. Data come in many forms. Some data are not numbers at all. Some data, called *categorical* data, only identify a category for an object or observation. For example, there are two categories for gender, male and female. Some categories for movie ratings are G, PG, PG-13, and R.

For some other data, a numerical value can take on any value on the real-number line. This is called *measurement* data. Measurement data use appropriate units to measure an observation or outcome. The time in minutes it takes you to travel to school is an example of measurement data. The amount of money required to make a movie is measurement data.

OBJECTIVES
Recognize differences between categorical and measurement data.
Use appropriate summary techniques for each type of data.

INVESTIGATE

Sometimes categorical data can be ordered or ranked. A restaurant rating of Excellent, Good, Fair, or Poor is an example of ordered categories.

Measurement data have numerical meaning, and it may make sense to add them, subtract them, and find averages. These data are usually summarized with a *measure of center* and a *measure of spread*. The measure of spread indicates variability.

Categorical data can be summarized by:

> Counts
> Percents
> Modal categories
> Bar graphs

Measurement data can be summarized by:

Median
Mean
Range
Quartiles
Number-line plots
Plots over time
Box plots
Scatter plots
Histograms
Stem-and-leaf plots

Discussion and Practice

1. Each heading below can be used to describe a data set. Choose at least five headings to discuss. For each one:

   a. Identify at least three items that could be in the data set.

   b. Indicate what method you would prefer to use to summarize each data set.

Ice cream flavors	Automobile colors
Gender	Race
Dominant hand	Citizenship
Last names	Letter grades
Movie ratings	Temperature
Age	Weight
Height	Blood pressure
Distance	Family size
Travel time	Grade-point average
Colors of shirts worn today	Elevation

Practice and Applications

2. Collect data to answer each question and write an appropriate data summary.

   a. What are the first three digits of the telephone numbers of students in your class?

   b. What time do students in your class get up in the morning on a school day?

   c. What was the last movie seen at a movie theater by students in your class?

   d. Which television station is watched most often on a given day?

3. List five questions you could ask in a survey that require different kinds of data.

In order to work with data using a calculator or computer, we often use numbers to represent all variables, even some variables that are categorical. For example, gender is sometimes coded 1 for male and 2 for female; and letter codes are sometimes given values like 4 for A, 3 for B, 2 for C, and so on. In other instances, data that were originally measurement data may be grouped as categories. For example, a category called "twenty-somethings" may be used to collect ages such as 23, 27, 29, and so on. "Teens" describes those from ages 13 to 19.

4. Consider the following numbers:

5, 23, 34, 51, 52, 66, 67, 73, 74, 81, 82

Suppose all of these numbers are in use in each of the situations below. Summarize the data in appropriate ways.

a. The numbers represent ages of hospital patients on a certain day.

b. The numbers represent daily temperatures on a sample of 11 days during the year.

c. The numbers represent football-jersey numbers for the Pittsburgh Steelers. Hint: 1–19 are used for quarterbacks only; 20–39 are used for running backs; 50–59 are used for centers; 60–69 are used for guards; 70–79 are used for tackles; and 80–89 are used for ends.

Summary

Different types of data values are appropriate for different problems. Categorical data represent the number of objects or observations that fall in clearly defined categories. Measurement data can be represented by any values on the real number line. Data may be sorted, counted, ordered, and summarized.

Developing an Investigation Process

Do any students in your class share the same birthday?

What do you think is the most popular month for birthdays?

INVESTIGATE

Birthdays of famous people are often reported in the newspaper or on the radio; they may also be celebrated as holidays. For instance, Dr. Martin Luther King was born January 15, 1929. This information may be written using numbers, 1-15-29. The context of the problem is important when using such data shortcuts. Only the fact that the discussion is about birthdays makes it clear that 1-15-29 is not a locker combination.

In many countries around the world, dates are written with the day of the month first.

Emily was born September 18, 1978. In some countries, this is written symbolically as 18-9-78. How would you write your birth date using this symbolism?

On which days of the year would you be confused about the date if you were not certain whether the month or day came first? How would you write twenty years from today's date in symbols?

Discussion and Practice

Birthdays are special days. Is today the birthday of someone in your class? Is someone having a birthday this week? This month? How many people have birthdays in the summer?

Many problems can be approached by collecting data and analyzing the results. Below is an outline of an approach that you will use in solving such problems.

Investigation Process

Problem: Identify the problem, usually by stating a question that needs to be answered.

Data Collection: Collect data needed to answer the question.

Data Presentation: Use tables, graphs, and charts to give a visual summary of your data. Also calculate any appropriate numerical summaries such as measures of center or spread.

Conclusion: Write a paragraph describing your conclusions about the answer to the question. You may also highlight things that your graph shows to be true, such as the meanings of clusters and gaps, the center and spread of the data, and so on.

The problem for investigation in this lesson is this: **Are certain months or dates more "popular" for birthdays than others?**

1. As a class, record every person's birthday. This is the *data set* that will be needed to complete the following investigations. As you will see, this data set will be used to find the solution to the problem.

2. Use your data to answer the following questions.

   a. On which days of the month were students born?

   b. Were more students born on even days or odd days?

   c. Which day of the month is the most common for birthdays?

3. Can your data be used to determine which days of the week students were born? Explain.

4. Can these data be used to determine how many students were born on the first day of the month? The last day of the month? Explain.

5. It is often helpful to create plots to identify patterns in data.

   a. Plot the frequency for each month in the data set.

   b. Plot days of the month students were born using a stem-and-leaf plot.

   c. Summarize each of the above plots in words.

Practice and Applications

When you presented your data, you probably arranged the months or days in order. When data can be ordered, it is often helpful to do so. A reasonable pattern for ordering birthday data is to arrange the months in order from January to December and the days in increasing order for each month.

Data that can be ordered may also be grouped. Group your data to answer the following.

6. How many students were born in the summer?

7. How many students were born in January or February?

8. How many students were born on or before the 15th day of any month?

9. Write a short paragraph that summarizes your findings based on any patterns you noticed in the collected data.

Summary

Data are often used to answer questions and support arguments. The investigation process outlined in this lesson provides a structure for using data in problem solving. It includes:

- Ask a question that identifies a problem.

- Collect data to answer the question.

- Study the data for patterns. Create appropriate tables, graphs, and plots. Graphs and plots may include number-line plots, box plots, stem-and-leaf plots, histograms, scatter plots, and bar graphs.

- Write a summary that describes your conclusions about the answer to the question.

Using Data in the Investigation Process

Have you ever wondered how many M&Ms there are in a small package?

Do certain colors appear more frequently?

Which colors do you prefer?

When you eat the M&Ms, do you group them by color?

In this lesson, you will use the investigation approach you learned in the previous lesson to look at packages of M&Ms and solve a problem.

INVESTIGATE

Based on previous experiences of eating M&Ms, would you expect the total number of M&Ms in a small package to vary much? Would you expect the number of one color to vary by very much?

OBJECTIVES

Use an investigation approach to solve a problem.

Count, sort, and display data to answer questions.

Discussion and Practice

Work together in a small group. Copy the table below or use *Activity Sheet 2*. Each group should have a small package of M&Ms.

M&Ms Color Data

Color	Individual Estimate of Frequency	Actual Frequency	Class Mean Frequency	Difference (Actual Frequency minus Class Mean Frequency)
_____	_____	_____	_____	_____
_____	_____	_____	_____	_____
_____	_____	_____	_____	_____
_____	_____	_____	_____	_____
_____	_____	_____	_____	_____
_____	_____	_____	_____	_____
Totals	_____	_____	_____	_____

Closeness score for actual counts _____

Closeness score for estimates _____

Problem: How many M&Ms of each color would you expect to find in a small package? Specifically, how many green M&Ms would you expect?

1. Before you open your package of M&Ms, complete the "Individual Estimate of Frequency" column in your table by guessing how many of each color of M&Ms and the total number of M&Ms are in your package.

2. *Data Collection*

   a. Sort and count the number of M&Ms for each color and record the information in the "Actual Frequency" column.

   b. Share the information with your classmates. Find the class mean, or average, for the frequency of each color and complete the "Class Mean Frequency" column.

   c. Complete the "Difference (Actual Frequency minus Class Mean Frequency)" column.

3. *Data Presentation*

   Create an appropriate plot for the data.

4. *Conclusion*

   Write a paragraph describing how many M&Ms of each color you would expect to find in a package. This will be referred to as the "average package" for the class.

5. Find out how close your counts are to the average counts.

   a. Use the difference column on your table to figure out a number that indicates a closeness score for your data. A closeness score should summarize the differences between your package and the average package. Identify the package closest to the class average.

   b. Create a graphical presentation for the closeness of your color counts to the class averages.

   c. How many small packages would you expect to open to collect a total of at least 25 green M&Ms? Will this number of packages always give you 25 green M&Ms? Explain.

6. In Problem 5, you used a closeness score to compare packages of M&Ms with the average package. Use the same closeness score procedure to determine which group in the class had estimates closest to the average package.

Practice and Applications

7. Now try another investigation involving your package of M&Ms. The problem is: About how much does one M&M weigh?

   a. Estimate how much one M&M weighs.

   b. Use the weight and number of M&Ms in an entire package to find an average weight of one M&M for each package. Share the information with other groups. Record the information for each package.

   c. Create a plot to show the variability in the estimated weights from one package to another.

   d. Write a summary of the results.

Summary

The investigation process includes:

- Identification of the problem

- Data collection

- Data presentation

- A summary of results

The summary may include information about the center of the data and how close the data are to the center. A measure of closeness of the data to the center indicates the amount of variability in the data.

Summary for Unit II

- Different types of data values are appropriate for different problems. *Categorical* data represent the number of objects or observations that fall in clearly defined categories. *Measurement* data can be represented by any values on the real-number line. Data may be sorted, counted, ordered, and summarized.

- Data are often used to answer questions and support arguments. The investigation process outlined in this section provides a structure for using data in problem solving. The investigation process includes:
  - Identification of the problem
  - Data collection
  - Data presentation
  - A summary of results

- Appropriate plots may include number-line plots, box plots, stem-and-leaf plots, histograms, scatter plots, and bar graphs. The summary may include information about the center of the data and how close the data are to the center. A measure of closeness of the data to the center indicates the amount of variability in the data.

What If Water Cost as Much as Cola?

Has water use ever been limited in your city?

What steps can you take to conserve water?

When you use water, do you think about how much it costs?

Gasoline and fuel oil were once thought so plentiful and inexpensive that people did little to conserve them. Times have changed, however, and today people realize that gasoline and oil are limited resources and should be used efficiently. In certain parts of the country today, water is still regarded as plentiful. In those areas, few people adjust their daily activities to conserve water.

INVESTIGATE

About how much water do you think you use in a day? About how much do you think each gallon of water costs?

Discussion and Practice

For a period of five consecutive days, each time you perform one of the activities listed on *Activity Sheet 3*, record a mark in the "Tally" column. Remember to include tally marks for activities that were performed for you such as washing clothes. If there are four people in your family and approximately $\frac{1}{4}$ of the clothes in the washing machine are yours, you should record 7.5 gallons for each use. Then complete the "Sum of Tallies" column. The totals should approximate the amount of water you used for those days.

Use *completed Activity Sheet 3* for the following problems.

1. Rank the activities according to the amount of water in gallons for each use.

2. As a class or in a group, choose five of the activities listed on *Activity Sheet 3* and complete a table like the one below, also on *Activity Sheet 4*, for your choices. Using the information on the table, write several sentences comparing water use for these five activities for a total of five days.

Data Summary of Water Usage

Activity	Number of Gallons You Used in 5 Days	Average Number of Gallons Class or Group Used in 5 Days	Difference: *Amount You Used* Minus the *Average Amount Used*
___	___	___	___
___	___	___	___
___	___	___	___
___	___	___	___
___	___	___	___

3. Make a graph showing your individual water usage for each day of the investigation.

4. Calculate how much water you would use in a week, a month, and a year.

5. Sometimes water is measured in cubic feet. One gallon contains 231 cubic inches.

   a. About how many gallons are in a cubic foot of water?

   b. How many classrooms could be filled up with the amount of water used by your class in a year?

6. In a midwestern city in the United States, the average price for water in 1997 was $2.54 per 1,000 gallons of water. Use this information to calculate about how much your family would spend for water use each week, month, and year.

7. Categorize water usage by rooms in your house. Where is water used the most?

One everyday liquid whose cost you may be familiar with is your favorite cola or other soft drink. Review soft-drink costs in Lesson 3.

Fill in the "Price per Gallon" column in your table for that lesson with an estimate of the price of a gallon of cola in your area.

8. Imagine that water cost as much as cola. How much would you spend in a month and in a year?

Practice and Applications

9. Find a person in your class whose water usage was different from yours. Compare your tables and write a paragraph explaining

    a. why your classmate's totals are different from yours.

    b. what changes you or your classmate could make to conserve water.

10. Write a report ranking and comparing water use for students in your class.

Using Percents

Percents and Relative Frequency

In a bag of 50 M&Ms, 4 are green and 10 are blue. What percent are green? What percent are blue?

Frequency indicates the number of times something occurs. *Relative frequency* provides a number useful for comparisons because it is a proportional measure obtained by dividing the frequency by the total number of possible occurrences.

$$\text{relative frequency} = \frac{\text{frequency}}{\text{number of possible occurrences}}$$

INVESTIGATE

Suppose a student has 50 M&Ms, including 4 green ones. Another student has 60 M&Ms, including 5 green ones. How can you compare the number of green M&Ms each student has? The first student has a relative frequency of $\frac{4}{50}$, or 8%, while the second student has a relative frequency of $\frac{5}{60}$, or about 8.3%. Relative frequency may be written as a fraction, a decimal, or a percent. What are some advantages of expressing relative frequency as a percent? When might it be better to leave it as a fraction?

Discussion and Practice

1. Is it possible for two packages of M&Ms to have the same frequency for one color but different relative frequencies for this color? Explain.

2. Describe a situation in which the number of yellow M&Ms in one bag is greater than a second bag, yet the second bag has greater relative frequency of yellow M&Ms.

Practice and Applications

Use the investigation procedure outlined on page 25 and the information you collected in Lesson 6 to answer this question:

Does it appear that the M&M/Mars Candy Company, producers of M&Ms, has a formula in percents for the color mix in M&M packages? What might the formula be for the color mix?

3. You will need the data collected in Lesson 6 on *Activity Sheet 2* to complete this problem.

   a. Copy and complete the table shown below, or use the table on *Activity Sheet 5*. Use your data from Lesson 6 for the first three columns.

M&M Relative Frequency Color Data

	Individual Package Data			Class Data	
Color	Estimated Frequency	Actual Frequency	Relative Frequency	Actual Frequency	Relative Frequency
————	————	————	————	————	————
————	————	————	————	————	————
————	————	————	————	————	————
————	————	————	————	————	————
————	————	————	————	————	————
————	————	————	————	————	————
Totals	————	————	————	————	————

   b. Based on your calculation of relative frequency for the class data, write estimates of the percent of each color in a typical mix of M&Ms.

   c. How much variability might there be in your estimates for part b? How could you make your estimates better?

4. Is it possible for two packages of M&Ms to have the same relative frequency for one color but a different frequency for this color? Explain.

5. Suppose a package of M&Ms contains 100 candies.

   a. How many candies of each color would you expect to find?

   b. The 10-by-10 grid on *Activity Sheet 5* represents 100 candies in an M&Ms package. Color the grid to represent your predictions for the number of candies of each color.

6. One M&M weighs about 0.03 ounce. About how many green M&Ms would you expect to find in a 1-pound package?

Extension

7. Obtain a 1-pound package of M&Ms.

   a. Predict the frequency for each color.

   b. Divide the M&Ms in the package among the students in your class to sort and count the colors. Compile the data.

   c. Write a paragraph explaining how these data compare to the data for small packages.

Summary

Frequency is the number of times something occurs. *Relative frequency* is a proportional measure obtained by dividing the frequency by the total number of possible occurrences.

$$\text{relative frequency} = \frac{\text{frequency}}{\text{number of possible occurrences}}$$

Relative frequency may be written as a fraction, a decimal, or a percent.

Percents and Visual Comparisons

How many people live in the United States?

How many people live in the world?

OBJECTIVES

Use percents to make comparisons.

Use grids to clarify comparisons.

In the history of the world it took until 1830 for the world population to reach 1 billion people. The population doubled to 2 billion by 1930. It took only 30 years, until 1960, to reach 3 billion. By 1975, the 4 billion mark was reached. Now the total is over 5 billion and still increasing.

INVESTIGATE

When a population uses up its resources, severe problems develop. Several such problems are starvation, illness, and depletion of forests. Can you think of other serious problems?

Discussion and Practice

1. Which areas of the world do you think are the most heavily populated?

2. Why do you think it may be important to study population trends?

This lesson will help you look at how populations are distributed on the continents. The population (in thousands) and the area (in thousands of square miles) for each continent are listed in the table on page 41.

Area and Population Data for the Earth

Continent	Area (1,000 sq. mi)	1990 Population (1,000s)
North America	9,400	277,000
South America, Latin America, Caribbean	6,900	450,000
Europe	3,800	499,000
Asia	17,400	3,286,000
Africa	11,700	795,000
Australia	3,300	26,000
Antarctica	5,400	Uninhabited
Entire World	57,900	5,333,000

Source: 1990 *World Almanac*

3. Refer to the table above.

   a. What are the units for area?

   b. Write a number that represents the area of North America in square miles.

4. Use the table on *Activity Sheet 6*. The activity sheet has the table above but includes two additional columns.

   a. What percent of the area of the world's continents is represented by the area of North America?

   b. Complete the "Area" column in the table for the rest of the continents.

   c. What percent of the population of the world is represented by the population of North America?

   d. Complete the "Population" column in the table for the rest of the continents.

5. When you add the numbers in the "Percent" column, is the sum exactly 100%? If not, why not?

6. Use *Activity Sheet 6* for this problem.

   a. On the grid labeled "Area," color in the percent of the area of the world's surface occupied by each continent. Use a different color for each continent. Label each area or provide a key.

   b. On the grid labeled "Population," color in the percent of the world's population represented by each continent. Use the same color choices as in part a. Label each area or provide a key.

7. Write a paragraph that compares the amount of land each continent occupies with the size of the population.

Practice and Applications

8. Refer to the table on *Activity Sheet 6*.

    a. List the continents in descending order by area.

    b. List the continents in descending order by population.

9. Use the first two grids on *Activity Sheet 7*, one to show areas and one to show populations of Canada and the United States. Use the data given below.

Area and Population of Canada and the U.S.

Country	Area (1,000 sq. mi)	1990 Population (1,000s)
Canada	3,849.7	28,114
United States	3,618.8	248,710

Extension

10. Use the third grid on *Activity Sheet 7*.

    a. Use the percent data for area and for population of the continents to make a scatter plot with the percent of Earth's land area on the horizontal axis and the percent of Earth's population on the vertical axis.

    b. Draw the line $y = x$ and use it to make comparisons.

    c. Write a short paragraph describing what the plot shows about the relationship between area and population.

Summary

Percent means *per hundred*. To find percent, divide the part by the total and multiply by 100. Percents can be used to compare data that are different, such as area and population.

Percents and Surveys

How would you answer the question "Would you prefer a hamburger you ordered to be packaged in a paper wrapper, a foil wrapper, or a box?"

As environmental problems become more challenging, questions about how to package products are often discussed. One question consumers may consider is whether they prefer to have their groceries packed in paper or plastic bags. Polls are often conducted to help manufacturers and salespeople learn what customers prefer.

Read the following article.

OBJECTIVE

Interpret the results of a poll using percents.

Paper or plastic: Who picks each

Cox News Service
Washington, D.C.—

A Gallup Poll has found that 48% of American consumers answer "paper" and 37% say plastic when a grocery store checkout clerk asks what kind of bag they want.

The polling organization phoned 1,021 adults to find the answer to this question. It said the percentages were reliable to within plus or minus 3 percentage points.

It found that 83% of the respondents said their stores offered the choice of bag types. About 10% said their grocers offered paper bags only. Only plastic bags were available for about 7% of the respondents.

About one in eight grocery shoppers—13%—use both paper and plastic bags, the poll found. Fewer than 2% gave an answer of anything other than plastic or both.

The study also found that 53% of men use paper bags as compared with 43% of women.

Shoppers under the age of 55 and those with incomes of $25,000 or more also were more likely to prefer paper.

The survey was commissioned by Kraft & Packaging Papers Division of the American Paper Institute, which represents companies that make brown paper bags. In the poll, most of the paper bag users said they regarded the containers as better for the environment. The survey found that 90% reuse the paper bags around their homes.

But there are environmental problems with both types of bags, Jan Beyea, an official of the Audubon Society, said in an article in the Audubon Activist.

He wrote: "Heretical as it may sound, some uses of virgin paper can be more damaging to wildlife than plastic substitutes. Papermaking pollutes the water, releases dioxin, contributes to acid rain and costs trees."

Plastics are no environmental bargain, Beyea also said. Plastics contaminate the oceans, he said, "they degrade very slowly, they are non-renewable and their production results in pollution."

He urged greater reliance on reusable carriers.

Source: *Milwaukee Journal*, May 3, 1992

INVESTIGATE

After reading the article above, what factors do you think should be considered when you decide whether to use paper or plastic bags?

How would you respond if a checkout clerk at a grocery store were to ask whether you prefer paper or plastic bags? Why?

Discussion and Practice

1. Poll the students in your class. Record one vote for each student in a table like the one shown below.

Class Preferences for Paper or Plastic

	Paper	Plastic	Both	Other	Total
Females	————	————	————	————	————
Males	————	————	————	————	————
Totals	————	————	————	————	————

2. Set up a table like the one that follows.

Paper and Plastic Preference Percents and Counts

	Paper	Plastic	Other	Total
Percent	48%	37%	————	100%
Number Making This Choice	————	————	————	————

a. In your table, record the results described in the Gallup Poll article.

b. Examine the counts in the two tables you completed. Are the results similar? Why or why not?

c. What percent of the males in the class prefer paper bags? What percent of the females in the class prefer paper bags? How do these numbers compare with the information in the article?

3. Write at least three questions you think were asked in the survey discussed in the article.

4. The Gallup Poll discussed in the article was conducted by telephone. Describe other ways data on using paper or plastic could be collected.

Practice and Applications

5. Refer to the Gallup Poll article.

   a. The article states that the results were reliable to within plus or minus 3 percentage points. Explain what this means.

   b. According to the article, which groups of people are more likely to use paper bags?

6. At a neighborhood grocery store in Phoenix, Arizona, about 11 bales of paper bags and 7 boxes of plastic bags are used each week. There are 500 paper bags to a bale and 1,000 plastic bags to a box.

   a. What is the total number of bags used each week?

   b. What percent of the bags used are paper?

   c. What percent of the bags used are plastic?

Extension

7. *The Universal Almanac,* 1994, reports that in 1991 there were 123,421,000 males and 129,257,000 females in the United States.

   a. Convert these numbers to percent of the total population of the U.S. in 1991. What percent of the total population in 1991 was male? Female?

   b. Assuming that these percents were the same for the 1021 surveyed in the article, about how many of the people surveyed were male? Female?

Summary

Polls are conducted to answer questions about populations by collecting data from a random sample of a larger population. Results within a stated range of percentage points from a poll may be a reliable prediction of what could be obtained by surveying the whole population. In most problems, surveying the whole population is impractical, so polls are widely used to obtain approximate results. Percents are used to compare samples and populations of different sizes.

Percents and Diagrams

What kinds of data are used to rate athletic teams or players?

Data recorded for basketball players include field-goal-completion percent; free-throw-completion percent; time played; scoring average; and number of assists, rebounds, steals, and turnovers.

INVESTIGATE

In the 1979–1980 season, the NBA instituted the 3-point-shot rule, and the NCAA followed suit in the 1986–1987 season. High schools also have a 3-point curve. Why are shots made from outside the large curve awarded 3 points instead of 2?

Sometimes data may be displayed in a diagram. Shown on the next two pages are the shot charts of four high-school basketball players: guards Eric and Jamie, forward Otto, and center Brian. These diagrams represent the players' performances in four different games. Each of these players attempted baskets, made a few, and missed a few. They significantly contributed to the final outcome of each game. To interpret the data represented by the charts, you need to keep these general concepts in mind.

- Shots are recorded for the first half of a game at one end of the chart and for the second half at the other end of the chart.

- An open circle on the shot chart indicates that the attempted shot was a miss; a circle with an × through it indicates the shot was made.

- A basket is worth 3 points if the ball is shot from outside the large curve and 2 points if the ball is shot from within the large curve.

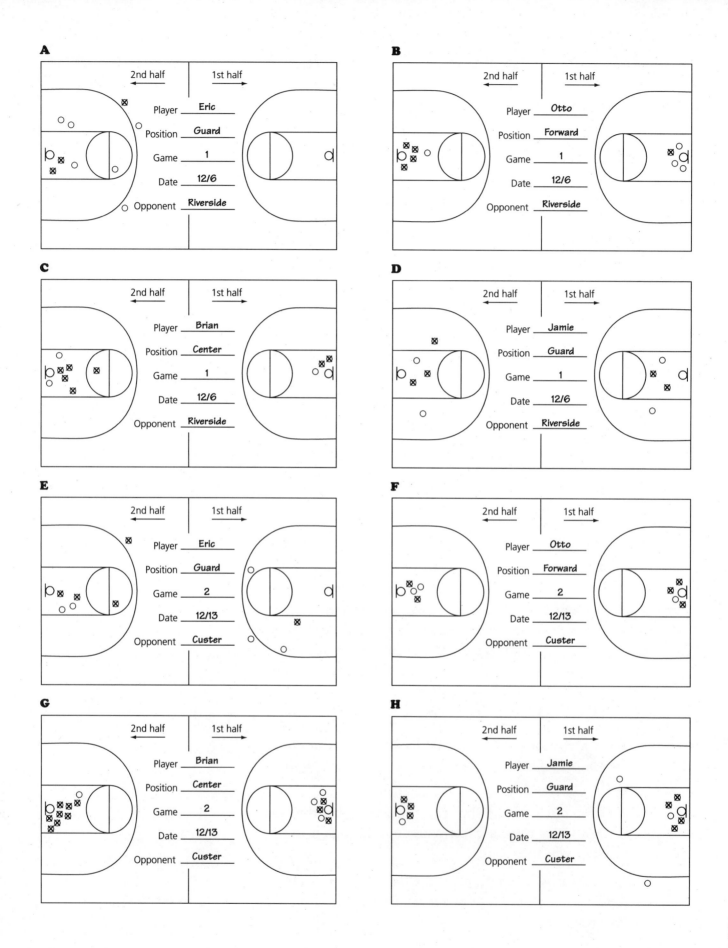

A

	2nd half ←	1st half →
Player	Eric	
Position	Guard	
Game	1	
Date	12/6	
Opponent	Riverside	

B

	2nd half ←	1st half →
Player	Otto	
Position	Forward	
Game	1	
Date	12/6	
Opponent	Riverside	

C

	2nd half ←	1st half →
Player	Brian	
Position	Center	
Game	1	
Date	12/6	
Opponent	Riverside	

D

	2nd half ←	1st half →
Player	Jamie	
Position	Guard	
Game	1	
Date	12/6	
Opponent	Riverside	

E

	2nd half ←	1st half →
Player	Eric	
Position	Guard	
Game	2	
Date	12/13	
Opponent	Custer	

F

	2nd half ←	1st half →
Player	Otto	
Position	Forward	
Game	2	
Date	12/13	
Opponent	Custer	

G

	2nd half ←	1st half →
Player	Brian	
Position	Center	
Game	2	
Date	12/13	
Opponent	Custer	

H

	2nd half ←	1st half →
Player	Jamie	
Position	Guard	
Game	2	
Date	12/13	
Opponent	Custer	

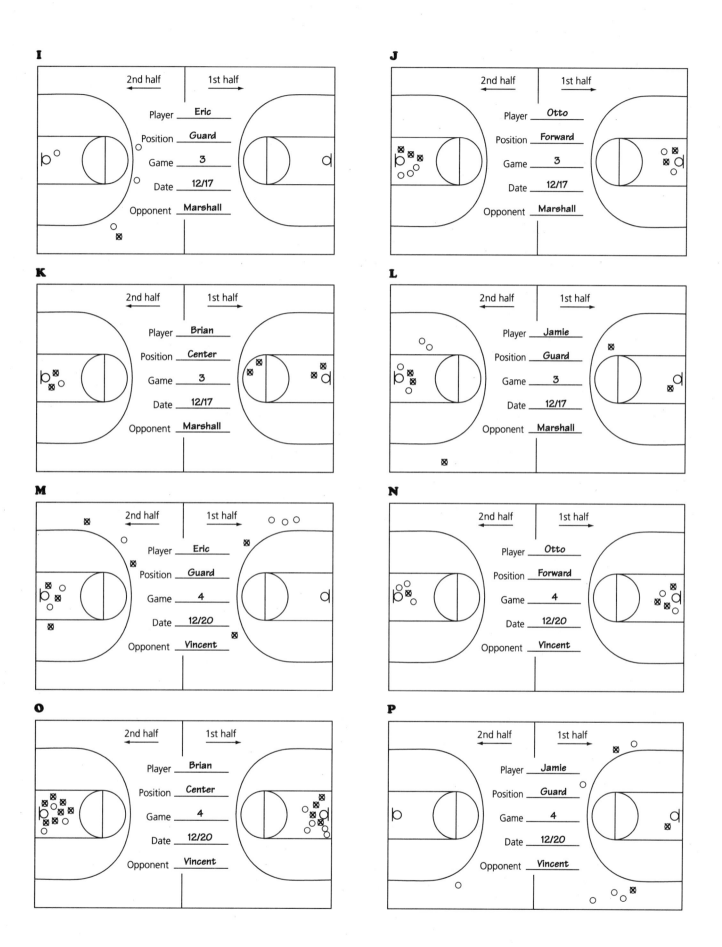

I

2nd half ← | → 1st half

Player _Eric_
Position _Guard_
Game _3_
Date _12/17_
Opponent _Marshall_

J

2nd half ← | → 1st half

Player _Otto_
Position _Forward_
Game _3_
Date _12/17_
Opponent _Marshall_

K

2nd half ← | → 1st half

Player _Brian_
Position _Center_
Game _3_
Date _12/17_
Opponent _Marshall_

L

2nd half ← | → 1st half

Player _Jamie_
Position _Guard_
Game _3_
Date _12/17_
Opponent _Marshall_

M

2nd half ← | → 1st half

Player _Eric_
Position _Guard_
Game _4_
Date _12/20_
Opponent _Vincent_

N

2nd half ← | → 1st half

Player _Otto_
Position _Forward_
Game _4_
Date _12/20_
Opponent _Vincent_

O

2nd half ← | → 1st half

Player _Brian_
Position _Center_
Game _4_
Date _12/20_
Opponent _Vincent_

P

2nd half ← | → 1st half

Player _Jamie_
Position _Guard_
Game _4_
Date _12/20_
Opponent _Vincent_

Discussion and Practice

1. Look at Shooting Charts A and E.

   a. How many 2-point shots did Eric attempt during the second half of game 1? How many 2-point shots did he make in the second half?

   b. How many points did Eric score in game 2?

2. Look at Shooting Charts F and J.

   a. How many 3-point shots did Otto attempt in game 2?

   b. How many points did Otto score in game 3?

3. Look at Shooting Charts O and P.

   a. How many shots did Brian attempt in game 4?

   b. How many 3-point shots did Jamie attempt in the first half of game 4?

In order to evaluate the performance of each player, it is helpful to organize the data.

4. For each game, complete a table like the following to organize and summarize the data given in the shooting charts. You may use the tables provided on *Activity Sheet 8*.

Player	Position	2-Point Shots Attempted	Completed	3-Point Shots Attempted	Completed	Percent Completed
Eric	Guard	_____	_____	_____	_____	_____
Otto	Forward	_____	_____	_____	_____	_____
Brian	Center	_____	_____	_____	_____	_____
Jamie	Guard	_____	_____	_____	_____	_____

5. Use your tables from the four games for the following problems.

   a. In terms of percent of shots that were made, identify the best game for each player.

   b. Create an appropriate plot of the percents of shots made for all four players using the data for all four games.

c. Which player would you say is the best player? Justify your answer.

   d. Which player(s) seem to be the most consistent? Justify your answer.

Practice and Applications

6. What percent of his 3-point shots in all four games did Eric make?

7. Draw a shot chart for a player who scored 14 points in a game. Use the diagrams on *Activity Sheet 9*.

8. Draw a shot chart for a player who scored 15 points in a game and had a shooting percent in the range of 30%–50%. Use the diagrams on *Activity Sheet 9*.

9. Another measure of success for basketball players is the average number of points scored per game.

   a. Find the average number of points scored per game for each player.

   b. Rank the players according to average number of points scored per game.

10. How many points would Eric need to make in the next game to average 12 points per game?

11. Eric wants to raise his point average by 2 points in game 5.

   a. How many points will he have to make in game 5 to meet his goal?

   b. How likely is his point average to increase by 2 points after game 5? Explain.

Extension

What role does distance play in the evaluation of a basketball player?

12. Examine Eric's shooting chart carefully and think about the relationship between his distance from the basket and his shooting percent. Write a short summary of your thoughts.

13. Examine Eric's and Otto's completion percents more closely by completing a table like the following or use *Activity Sheet 10*.

Completion Percents

Player	Position	Percent Completed			
		Game 1	Game 2	Game 3	Game 4
Eric	Guard	————	————	————	————
Otto	Forward	————	————	————	————

14. Use the tables labeled "Eric's Shots" and "Otto's Shots" on *Activity Sheet 10*.

   a. Measure the distances from the basket in centimeters or millimeters for each of Eric's and Otto's shots in each game.

   b. Find the average distance from the basket for all 2-point shots in each game.

   c. Find the average distance from the basket for all 3-point shots in each game.

   d. Record the data in the tables on the activity sheet.

15. Use the data from the tables in Problem 14 and represent the data in at least one plot or graph.

16. Examine the data for Eric and Otto (Problems 14 and 15) and describe the role you think distance from the basket plays in their shooting performance.

Summary

Diagrams and tables may be used to organize data and look for trends, patterns, and relationships. Percents may be used to make comparisons.

Percents and Probability

If the probability of rain were 60%, would you carry an umbrella?

If the probability of winning a game were 0.15, would you expect to win?

The *probability* of an event is a measure of how likely it is that the event will occur. Probability is expressed as a value from 0 to 1. A probability of 0 indicates that an event will not occur. A probability of 1 indicates that an event is certain to occur.

INVESTIGATE

The students in Homeroom 207 have decided to sponsor a game booth at the school's fund-raiser. They considered several ideas for games, but they voted to design a "Cup-Toss" game. "We'll have to know what the probabilities are for all of the possible outcomes," Jacqueline said. Jacqueline is the class treasurer. "We want to make money for the school, not give it away."

So the class got some cups and divided up into pairs to collect some data on what happens when a cup is tossed in the air. "I'll bet it lands every time," laughed Chris, the class clown.

"Yeah, and the probability is one that you will goof off every time we have work to do. It's a sure thing," Jacqueline came back.

Why did Jacqueline say that a probability of 1 is a sure thing?

Discussion and Practice

The table on page 54 shows the data collected by students in their experiment.

1745 Paper-Cup Tosses in Homeroom 207

Groups	Bottom	Top	Side	Totals
Group 1	7	22	171	200
Group 2	11	22	167	200
Group 3	12	32	306	350
Group 4	11	25	264	300
Group 5	10	49	241	300
Group 6	12	26	212	250
Group 7	2	10	34	46
Group 8	1	27	71	99
Totals	66	213	1466	1745

1. Examine the data that Homeroom 207 collected to complete the following problems.

   a. Which outcome seems most likely to occur? Which outcome seems least likely?

   b. Which group had the most cups land on the side? Which group had the least?

   c. Compare the results for groups 1 and 3.

2. Discuss rules for how a cup should be tossed in the Cup-Toss game.

Work with a partner to analyze the cup-toss data from Homeroom 207. Use a table like the one below, provided on *Activity Sheet 11,* or use a spreadsheet.

Paper-Cup Tosses

	Bottom		Top		Side		Totals
	Number	Percent	Number	Percent	Number	Percent	
Group 1	———	———	———	———	———	———	———
Group 2	———	———	———	———	———	———	———
Group 3	———	———	———	———	———	———	———
Group 4	———	———	———	———	———	———	———
Group 5	———	———	———	———	———	———	———
Group 6	———	———	———	———	———	———	———
Group 7	———	———	———	———	———	———	———
Group 8	———	———	———	———	———	———	———
Totals	———	———	———	———	———	———	———

3. Fill in the "Number" columns in the table. Then, for each group, show the outcomes as percents—comparing the number of times each outcome occurred to the total number of tosses that the group performed.

    a. Find the percent of the total trials that resulted in bottoms, in tops, and in sides.

    b. Does your percent table help you analyze the results better than the tally table? Explain.

4. The probability of an event is a measure of how likely it is to occur. Probability may be expressed as a fraction, a decimal, or a percent.

    a. Based on the experimental data collected in Homeroom 207, what is the probability that a paper cup tossed in the air will land on its side?

    b. What is the probability that a paper cup will land with the top side down?

    c. What is the probability that a paper cup will land with the bottom side down?

Practice and Applications

Profit or Loss?

Suppose the booth offers a $1 prize for a cup landing on its bottom and a $0.75 prize for landing on its top. There is no prize if a cup lands on its side.

5. Suppose the groups had charged $0.15 per toss.

    a. Would any of the groups have made money?

    b. Would the class have made a profit or lost money on the game? How much?

6. Make up a different fee and prize structure for the cup-toss booth. It should be one that you think would result in more money for the class. Show in a table how much you would expect to take in and pay out using your system.

The cup-toss experiment provides information about the true performance of the cup when it is tossed into the air. It also allows you to make reliable predictions about what will happen in the next 200 or 2,000 tosses. Divide your class into 8 groups to conduct your own experiment like that conducted by Homeroom 207.

7. Work with a partner or group to collect data.

   a. Toss a paper cup many times. Keep track of your results in table form and compare your data with the data collected by Homeroom 207.

   b. Based on your class data, were the results of Homeroom 207's experiment what you would expect? Support your claim.

8. The students in homeroom 207 did some additional experimenting to see if practice could make them better at the cup-toss game. Their data are shown below. Analyze the data. What results did they find in these experiments? Did they "get better at the game"?

Room 207 Data

Bottom	Top	Side	Totals
6	24	70	100
8	26	66	100
5	26	69	100

9. How do these new data change your original estimate of the probabilities for the cup-toss game?

Extension

10. Investigate what happens if you tamper with the cup. Figure out a way to change the weight or the size of the cup in such a way that you suspect the probabilities will be affected.

    a. Predict how your change will affect the probabilities of side, top, and bottom.

    b. Do the experiment several times with your group's tampered version of the cup and record your results in table form.

    c. Write a paragraph comparing your group's results with the original results or with the results of other groups.

Summary

Variation will occur in collecting data. A certain amount of variation is expected. *Probability* can be shown as a fraction, a decimal, or a percent. Probability may be used to predict the likelihood of an event. The value of a probability may be any number from 0 through 1.

Summary for Unit III

- *Frequency* is the number of times something occurs. *Relative frequency* is a proportional measure obtained by dividing the frequency by the total number of possible occurrences.

$$\text{relative frequency} = \frac{\text{frequency}}{\text{number of possible occurrences}}$$

  Relative frequency may be written as a fraction, a decimal, or a percent.

- Percents may be used to answer questions about data. *Percent* means *per hundred*. To find percent, divide the part by the whole. Percent is an appropriate measure to use to make comparisons.

- Polls are conducted to predict outcomes by collecting data from a random sample of a larger population. Data within a stated range of percentage points from a poll may be reliable as a predictor of the larger population. Percents are used to compare populations of different sizes.

- Diagrams and tables may be used to organize data and look for trends, patterns, and relationships. Percents may be used to make comparisons.

- *Variation* will occur in collecting data. A certain amount of variation is expected.

- *Probability* can be shown as a fraction, a decimal, or a percent. Probability may be used to predict the likelihood of an event.

- Data may be collected by making observations, conducting surveys, or performing experiments.

Assessment for Unit III

Your class has taken a mathematics test. The scores, reported as percents, are listed below.

83	78	87	90
93	79	57	85
99	81	85	78
86	64	83	100
86	92	93	86
70	83	83	95

Your score, included in the table, was 86%. Your assignment is to explain how well the class did as a whole and how well you did compared to the rest of the class. You should use the investigation process you learned in Unit II.

1. *Problem:* Write a sentence describing the problem.

2. *Data Collection:* The data are listed above. Complete parts a, b, and c so you can consider the center of the data and the variability in the data.

   a. Find the mean, or average, of the data to describe the "center" of the data.

   b. Copy and complete a table like the following for all of the scores.

Score	Difference (Score Minus Class Mean)
83	————
78	————

   c. Summarize the information in the difference column to describe the variability in the data set.

3. *Data Presentation:* Create an appropriate plot or graph of the data.

4. *Conclusion:* Write a paragraph that addresses the problem you identified. Justify your conclusion. Include information about the center and the variability of the scores.

5. For the following grading scale, find the percent of As, Bs, Cs, Ds, and Fs.

   A 90–100%

   B 80–89%

   C 70–79%

   D 60–69%

   F below 60%

Measurement Data in Experiments

Exploring Centers and Variability

What is the difference between the measure of center of the heights of the players on a basketball team and the height of the center?

The measure of center of the heights can be the mean or median of the heights of the players. It is a number that summarizes or averages the heights of the players. It is not necessary for any player on the team to have the center or average height. The height of the center is the height of the player that plays the center position.

Measurement data can take on any value on the real number line. Appropriate units are used to measure an observation, result, or outcome. It is often helpful to summarize measurement data by finding a center.

OBJECTIVE

Explore the concepts of center of measures, variability, and bias.

INVESTIGATE

Have you ever thought about how the weight of a cow is measured? Practically speaking, it is impossible to expect a cow to stand still on a scale while you read the dial. So, while a cow stands on a scale, several measures are recorded over a period of a few seconds and the average or center of the measures is determined to be the weight.

Sometimes measurement data can be biased. *Bias* shows favor toward one or more outcomes. The activities that follow show how some factors can affect measurement data. It is important to minimize bias in data collection. It may also be important to notice the variability or spread in a data set.

Discussion and Practice

1. Comment on how the word *center* is used in each of the following:

   Center of a circle
   Center of the universe
   Population center
   Center of mass
   Communication center
   Center of measures

2. Why would it be more reliable to average several weights rather than take the first one recorded when a cow stands on a scale?

3. Describe another situation where the center or average of repeated measures is used.

Practice and Applications

4. **Problem A** Estimate the height of the center of a doorknob chosen by the class. Half of the students should estimate the height in inches and the other half should estimate the height in centimeters.

   a. Write your estimate on a piece of paper and give it to your teacher or to a recorder. Record the results on the board or an overhead.

   b. Find a measure of center of the estimates in inches and a measure of center of the estimates in centimeters.

   c. Make a box plot for the inches data and a box plot for the centimeters data.

   d. Describe the variability within each data set.

   e. Measure the height of the center of the doorknob. Determine if one group estimated with greater accuracy than the other. If so, explain why that may have happened.

5. **Problem B** Estimate the length of a given segment using the figure provided on *Activity Sheets 12 and 13*.

   a. Develop and carry out a plan that will result in good data collection in which one student's estimate will not influence others.

   b. Make a box plot of the estimates.

c. Describe the variability in the estimates.

d. Measure the correct length of the segment and mark this measure with an X on your box plot.

e. Where does this X fall in relation to your box plot? Describe any factors that may have affected or biased your estimates.

6. Describe some ways that bias can enter into a data-collection procedure for a survey or an experiment.

7. The heights of the players on the Nicolet High School girls' basketball team are listed below.

**Nicolet High School
Girls' Basketball Team**

Number	Name	Height	Year
12	Beth	5'4"	9
20	Liz	5'6"	9
22	Jessie	5'6"	11
24	Corey	5'8"	12
32	Jessica	5'4"	12
34	Angela	5'4"	12
40	Kristin	5'10"	11
42	Caterricka	5'5"	9
44	Colleen	5'10"	9
52	Jill	5'10"	12
53	Shaundra	5'7"	12
54	Jennie	5'10"	12
55	Jennifer	6'2"	12

a. Find the average of the heights of the players on the team.

b. Which girl do you think plays center?

Summary

Continuous data are used to find measures such as length, area, volume, and time. Some variation is expected in collecting data. If there is more or less variability in a data set than is likely to have occurred, you may look for sources of *bias*. The mean and median may be used to describe the center of a set of measurement data.

Reaction-Time Experiment

How quickly do you react when faced with a stimulus?

When is a fast reaction time important?

Your *reaction time* is a measure of the time it takes you to respond to a stimulus. Do males react faster than females do? The experiment in this lesson studies reaction time by dropping a ruler through the fingers of a student and measuring how far the ruler falls before it is caught.

INVESTIGATE

Before doing this experiment, some decisions need to be made about what methods or techniques will be used to minimize experimental error. Read the experimental procedure below. Then try the procedure a few times and decide which dropping-and-catching techniques should be controlled. Consider these variables:

- How far above the hand should the ruler be held?
- How many fingers should be used to catch the ruler?
- Should the measurement be made above or below the fingers?

Are there other important variables to consider?

Experimental Procedure

Work in pairs to conduct this experiment.

- One of you holds a metric ruler and drops it through the open fingers of the other.
- Measure how far the ruler falls before it is caught.
- Each of you should complete the procedure ten times.
- Record the data in a table.

Discussion and Practice

Problem: How long does it take you to react when a ruler is dropped through your fingers? Do males in the class react faster than females do?

Data Collection

1. **a.** Perform the experiment ten times with one person catching the ruler. The person who drops the ruler will also record the distances the ruler fell each time.

   b. Reverse roles and perform the experiment ten more times.

   c. Write at least three observations about each set of ten drop distances.

Data Presentation

2. Plot the data sets separately in an appropriate way and calculate a center distance for each person using mean, median, or mode.

Pictured below is the graph "Time for a Freely Falling Body to Drop a Given Distance." Use it to convert your distance to a reaction time. For example, if the ruler fell through your fingers a distance of 20 cm before you caught it, the time it took you to react and catch the ruler was 0.2 second.

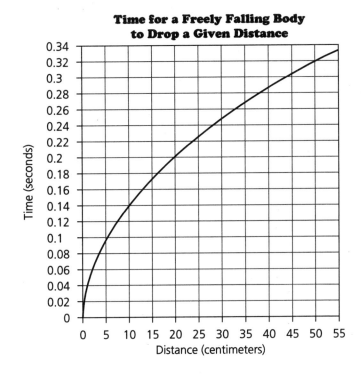

3. Convert the center distance to a measure of reaction time. Record it with "F" for female and "M" for male. Record all the results on the board or an overhead projector.

4. On one set of axes, make two box plots, one for female and one for male reaction-time data.

Conclusion

5. Write a paragraph about whether males appear to have faster or slower reaction times than females do.

Practice and Applications

6. Examine your own data a little more closely by making a different plot of your drop distances than the one you made in Problem 3. For example, if you made a plot over time, you could make a box plot or a stem-and-leaf plot.

   a. Describe any trends or patterns in your reaction-time data.

   b. Predict the results of ten or more tosses.

7. Describe the variability in your data set.

8. If this experiment were done with the catcher blindfolded and the dropper giving a verbal or tactile clue, do you think the results would be different? What would you record if the person missed the ruler entirely?

9. How do you think using your opposite hand would affect results? Explain.

Extension

10. The distance it takes a car to stop depends on the reaction time of the driver and the speed of the car. The reaction time is the interval between when the driver recognizes the car must stop and the brakes are applied. In the time it takes a driver with normal reaction time to react to an emergency, a car travels a distance of d feet, where $d = 1.1$ times the speed of the car. Find the distance in feet a car going the given speed travels before the driver hits the brakes.

   a. 35 mph   b. 65 mph

Summary

We conduct experiments to collect data to answer a question or solve a problem. In an experiment, variables that may affect results should be considered. Some variables may be controlled.

Burst-Your-Bubble Experiment

How large a bubble do you think you can blow with the "bubble stuff" you used as a young child?

OBJECTIVES

INVESTIGATE

The experiment below will help you answer the question "How large a bubble can you blow?" Estimate the diameter of a bubble you could blow using a straw. Decide as a class whether you would estimate in centimeters or inches. Share your estimates and find a reasonable class estimate for the size of a bubble.

OBJECTIVES

Collect experimental data.

Use measurements in the investigation process.

Discussion and Practice

Work with a partner. You will need bubble mixture, a straw, and a ruler. Take your books off your desk. If you don't want to get bubble stuff on your desk, cover it with paper towels or newspaper. Lay the ruler on your desk.

1. Decide how you might measure a bubble's diameter.

2. Put a straw in the bubble mixture and blow a bubble. Find the diameter of the sphere you have blown by reading the measurement from the ruler. You may have to do this quickly, perhaps through the bubble while it is still attached to the end of your straw.

3. Blow at least 5 bubbles and record their diameters.

Complete the investigation process.

Problem: What is the average diameter of the bubbles you can blow with a straw?

4. *Data Collection*

Collect data from students in your group. Each student should make at least 5 bubbles. Organize your data in a table or chart that can be turned in with your conclusions.

5. *Data Presentation*

a. Calculate a measure of center using mean or median.

b. Create an appropriate graph or plot of your data.

6. *Conclusion*

a. Summarize your results. Include a measure of center and a description of the variability in your data.

b. Explain any problems that you had or things you could have done to make your investigation easier.

Practice and Applications

7. Compare your group's answer to the class estimate.

8. Identify another problem or question that the investigation process could be used to answer.

9. The volume of your bubbles can be calculated using the formula volume $= \frac{4}{3}\pi r^3$, where r = radius, or $\frac{1}{2}$ the diameter. Find the volume of your average bubble.

Extension

10. Make a poster describing your experiment.

Summary

- The investigation process includes:
  - Identification of the problem
  - Data collection
  - Data presentation
  - A summary of results
- Tables, charts, graphs and plots may be used to organize experimental data.
- The summary may include information about the center of the data and the variability in the data.

Summary for Unit IV

- Continuous data are used to find measures such as length, area, volume, and time. Some variation is expected in collecting data. If there is more or less variability in a data set than is likely to have occurred, you may look for sources of *bias*. **Bias** shows favor toward one or more outcomes. The mean and median may be used to describe the center of a set of measurement data.

- Experiments are conducted to collect data to answer a question or solve a problem. Variables that may affect the results should be considered. Some variables may be controlled.

- The investigation process includes:
  - Identification of the problem
  - Data collection
  - Data presentation
  - A summary of results

- Tables, charts, graphs and plots may be used to organize experimental data.

- The summary may include information about the center of the data and the variability in the data.

Waiting Time in the Lunch Line

What is the longest time you have spent waiting in line?

_____

What are some things for which people wait in line?

_____

Americans spend hours waiting in line. Lines form at the checkout counter, the box office, the bank, the post office, the park, and the classroom door. People even wait in line on the telephone. Lines provide an orderly system for serving people. Sometimes statisticians study the amount of time customers must wait in line for service.

INVESTIGATE

One of the places in your school where you may have to wait in line is the lunchroom. The amount of time you wait in line may affect how long you have to eat or talk to your friends.

This project involves collecting data and analyzing it to find out how much time students spend waiting in the lunch line(s) at your school.

The problem you are to address is: **How much time do students spend waiting in the school lunch line?**

Discussion and Practice

1. Identify the variables related to the problem and develop a plan for collecting data.

   a. When should the data be collected and for how many days?

   b. Are the lines longer when certain foods are served?

c. What categories should be included? Are there different lunch lines and different lunch times? Will students who bring their lunches and don't stand in line at all be considered?

d. How should wait time be defined?

e. How should the data be collected? One possibility is to hand each student a card as students get in line. On the card, the specific line and a time indicating the beginning of the wait time is recorded. At the end of the waiting time, this card is handed back to a data collector and the end of waiting time is recorded. You might devise and use cards like the one shown below or cut out the cards on *Activity Sheet 14*.

```
┌─────────────────────────────────────────┐
│       Lunch-Line Waiting Time            │
│                                          │
│  Date      _____        │
│                                          │
│  Line      _____        │
│                                          │
│  Start Time_____        │
│                                          │
│  End Time _____         │
│                                          │
└─────────────────────────────────────────┘
```

2. Before beginning the collection of data,

   a. estimate the average amount of time you spend waiting in the lunch line on a school day.

   b. estimate the percent of your total lunch time you spend waiting in line.

3. Collect waiting-time data for at least two days. Set up a table or use the one on *Activity Sheet 15* to compile the data.

4. Summarize the data using the following procedure.

   a. Make at least one plot of the data.

   b. Describe any patterns you see in the data.

   c. Calculate the mean waiting time and the median waiting time for each day you collected data. Compare them.

   d. Find the range of waiting times.

5. Analyze the data.

   a. Convert the waiting-time data to percents of total lunch time students spend standing in line.

   b. Make a box plot of the percents.

   c. How do the actual percents compare to your estimate from Problem 2?

6. What is the "typical" daily waiting time in the lunch line for students in your school? Consider measures of center and variability in answering this question.

7. Do you think the amount of time scheduled for lunch is about right, too short, or too long? Use your results to support your answer.

8. Write a letter to the administration of your school that includes the results of your investigation.

Extension

9. Present your results to the administration or student government at your school.

10. Identify some other issues in your school that might be resolved by using a similar investigation process. An example is the amount of time scheduled between classes.